Images of America
Bethlehem Steel

This aerial view shows the Bethlehem Steel plant and the Lehigh River that separated the north and south sides of Bethlehem. City hall is in the foreground along with the Fahy Bridge, which connects to the South Side and the Bethlehem Steel plant. (Author's collection.)

ON THE COVER: A dedicated crew of machinists and material handlers from the manufacturing division pose for a photograph surrounded by rolls or shafts in the foreground. The large Ingersoll planer-milling machine at center did most of the flat machining in the No. 8 Machine Shop at the Bethlehem Steel Company's Bethlehem plant. (Courtesy of Steelworkers Archives.)

IMAGES of America
BETHLEHEM STEEL

Tracy L. Berger-Carmen

Copyright © 2022 by Tracy L. Berger-Carmen
ISBN 978-1-4671-0552-1

Published by Arcadia Publishing
Charleston, South Carolina

Printed in the United States of America

Library of Congress Control Number: 2020934630

For all general information, please contact Arcadia Publishing:
Telephone 843-853-2070
Fax 843-853-0044
E-mail sales@arcadiapublishing.com
For customer service and orders:
Toll-Free 1-888-313-2665

Visit us on the Internet at www.arcadiapublishing.com

*This book is dedicated to all the men and women
who worked in the steel industry.*

Contents

Acknowledgments		6
Introduction		7
1.	Outside Looking In	9
2.	Inside "The Steel"	17
3.	Off the Clock	53
4.	The People	63
5.	Products, Projects, and Tools	85
6.	Homer Labs, SGO, BPO, and Martin Tower	97
7.	Revitalization, Preservation, and Future Plans	119

Acknowledgments

I would like to thank the following people who helped in some way, gave me tidbits of information, provided computer support, or contributed photographs and stories: James Atherton, Frank Behun Sr., Fran P. Bush, Victoria Bastidas, James Benetzky, Carville Bevans, Martha Capwell-Fox, Nick Chismar, Kenneth Eberts, Al Fetterolf, Linda Finken, Nancy Frantz, Scott Gordon, John Hrkach, John Huber, Jodi Keim, Ron Keschl, Scott Krycia, Alexandra Lane, Mark Licisky, Robert Lightner, John Marquette, Jacob Matus, Joe Mayer, Rudy Miller, Conrad Milster, Ed Pany, Robert Penchick, Karen Samuels, Ken Sem, Bruce Ward, Donald Young, and Susan Vitez.

A huge thank-you goes out to the retirees of Steelworkers Archives, as well as the friends on the Facebook pages "Bethlehem Steelworkers and Their Families and Friends" and "Remembering Bethlehem Steel and Martin Tower–The Book," who answered questions, provided details, identified people, and contributed significantly to making this book a success.

Special thanks to the National Museum of Industrial History (Bethlehem, Pennsylvania), Delaware & Lehigh National Heritage Corridor (Easton, Pennsylvania), the Hagley Museum (Wilmington, Delaware), and the Bethlehem and Easton Area public libraries for letting me comb through thousands of pictures, articles, and documents, which I had to ultimately narrow down to 228 images.

A huge thank-you also goes to my husband, Larry, and mom, Diana Berger, who helped hold down the fort and watch the kids while I spent hours looking through slides and pictures, doing interviews, and traveling to get the pictures and information that I needed. It was truly a labor of love.

INTRODUCTION

The history of Bethlehem Steel goes back to 1857, when Sauconia Iron Company was formed. It became the Bethlehem Rolling Mill Company, followed by Bethlehem Iron Company, and finally what we know it as today, Bethlehem Steel Corporation.

It spanned for miles along the Lehigh River and Lehigh Canal in South Bethlehem, 1,600 acres total, which brought in raw product and shipped out finished steel. It made the steel that at one time made up 80 percent of Manhattan's skyline, along with many colossal bridges, some a mile long; it made ammunition and ships for the world wars, rails for the railroads, and everyday products for homes, as well as the Hadfield steel that couldn't be cut or filed that was used in prisons. This was Bethlehem Steel. This is what won the wars and essentially created modern America during the Industrial Revolution.

The company employed over 300,000 at one point, consisting of workers from generations of family members to entire households working for "The Steel." An entire city was created around it. Businesses thrived. People emigrated from all over the world with the promise of work during the darkest days of the Depression to the heyday of the 1950s and 1960s, when business was booming. Work was around the clock, and bonuses were given to top executives for a job well done. It was the largest steel-producing company in the United States and the second-largest in the world. Its plants, mines, and offices were global. Saying it was a huge operation is an understatement.

Then it all came down—slowly. The imports of foreign steel starting in the 1970s, the antiquated equipment in the plant, those generous bonuses to upper management, and superficial spending in other departments caused it all to collapse. The shops in the plant closed one at a time. The Martin Tower offices were the last to leave. In 2003, Bethlehem Steel's assets were dissolved. The last remaining six plants were acquired by Mittal Steel, which merged with Arcelor in 2005 to form ArcelorMittal.

Many buildings remain at the Bethlehem Steel site. Some date back over 100 years. Some have been torn down, some are being preserved, and some are sitting untouched, waiting for the next chapter.

One
OUTSIDE LOOKING IN

This 1873 map of Bethlehem shows some of the early manufacturing buildings that the Bethlehem Steel Company took over at the turn of the 20th century. Some of the landmarks on this map are still here today. The Hill to Hill Bridge, Fahy Bridge, and the Moravian church remain, as well as a few steel plant buildings. (Courtesy of Bethlehem Area Public Library Archives.)

Artist Joseph Pennell (1857–1926) painted this image in 1881. It is one of the earliest images in the archives of the Bethlehem Steel Works. It shows the stacks blowing fire and smoke from across the Lehigh River. The building on the right remains at the site but is currently deteriorating. (Author's collection.)

This 1917 postcard shows the blast furnaces along the Lehigh River. The river and the railroad were very important to Bethlehem Steel to ship supplies in and products out. It took four tons of water to make one ton of iron in a blast furnace and about 10 times that to make a ton of steel. (Author's collection.)

This 1905 postcard sent to Brooklyn, New York, shows the Bethlehem Steel Company, the Lehigh River, and the railroad tracks. The Holy Infancy Catholic Church to the right was adjacent to the Bethlehem Steel Company, and many of the steelworkers chose this church, as did many others in South Bethlehem. (Courtesy of Scott Gordon.)

The Lehigh Zinc Company was founded by Samuel Wetheril, who built zinc oxide furnaces along the Lehigh River. Mules and wagons would bring zinc ore from a mine in Freidensville four miles away. In 1853, Joseph Wharton bought interest in the Pennsylvania and Lehigh Zinc Company, and it became the first successful producer of zinc in the United States. Bethlehem Steel bought the property in 1911, and the company then moved to Palmerton, Pennsylvania, where it remained in operation until 1983. (Courtesy of Scott Gordon.)

This postcard shows the Alloy Division works, looking southeast from the New Street Bridge. The No. 3 open hearth is on the left with the 35-inch mill soaking pits in the center with their four stacks. During its peak, the company spanned over four miles along the river in South Bethlehem. (Photograph by G.A. Conradi, courtesy of Scott Gordon.)

Seen here are forge No. 2 as well as treatment buildings Nos. 7–9. The building remains on the Bethlehem Steel site, forging products for power plants and pressure vessels and supplying the oil, gas, defense, and metal industries. (Courtesy of Scott Gordon.)

In the early 1900s, there were many stacks throughout the Bethlehem Steel site. In an effort to make iron more efficiently, Charles Schwab took down many of those stacks. He was interested in producing wide flange steel beams. The old-fashioned I beams were replaced by the H beams designed by Henry Gray. (Courtesy of the Bethlehem Area Public Library.)

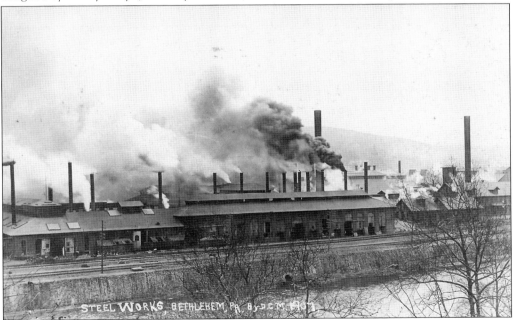

Smoke and steam billow from many buildings. Over the years, Bethlehem Steel Company faced multiple lawsuits over air, water, and soil pollution. The EPA was formed in 1970 and worked to ensure environmental protection as well as the health and safety of the residents in the area of the plant. (Photograph by G.A. Conradi, courtesy of Scott Gordon.)

This photograph was taken from atop the Steel General Offices (SGO) building. It was centrally located in the plant so that the upper management could efficiently oversee the operations. To the left is a machine shop beside the large overhead cranes used to move products in and out of the building. (Author's collection.)

The ore and limestone storage yards sat along the Minsi Trail Bridge. The gantry cranes scooped up the ore and put it in cars traveling on the Hoover-Mason Trestle to the blast furnace storage bins to begin the steelmaking journey. (Courtesy of Scott Gordon.)

This early 1900s postcard shows the alloy steel mill buildings off of Second Street, looking northwest. The far left building remains. It is now called Riverport, consisting of condominiums, a fitness facility, and a restaurant. (Courtesy of Scott Gordon.)

This 1911 postcard shows a partial panorama of the Bethlehem Steel corporation along Second Street with some landscaped gardens in the foreground. Bethlehem Steel had open accounts with many local businesses and spared no expense when it came to the beautification of the grounds. The 1873 Bessemer steel building to the left remains today as a shell. (Courtesy of Scott Gordon.)

This early 1900s postcard shows a boiler house with the coal elevating tower on the right. Railroad cars ran through the plant to drop off raw product and ship out the final product, providing efficiency to daily operations. (Courtesy of Scott Gordon.)

This aerial view of Third Street shows the shops that lined the road before the Minsi Trail Bridge. In the foreground is the roll foundry, in the middle is the steel foundry, and at left is open hearth No. 1 and the No. 2 machine shop. The tall building in the center at rear is the No. 2 machine shop annex. Some of the buildings have since been torn down to make room for the Wind Creek Casino, hotel, and outlet center. (Courtesy of Steelworkers Archives.)

Two

Inside "The Steel"

The pattern shop was built in 1913 at a cost of $5 million. It was a four-story brick and steel building along Second Street. It was put up rapidly by contractors Guerber Engineering and W.F. Danzer & Company to keep up with the demand for pattern production. (Courtesy of Scott Gordon.)

Crucible Steel workers pose for the photographer. Crucible Steel was initially started in England in 1790, but Charles Schwab wanted to upgrade the process since Bethlehem Steel was noted for its high-quality steel that was designed to cut steel with steel. A refractory crucible was charged with wrought iron, pig iron, and other alloys and poured into a coal-fired pit, which melted it down to high-carbon steel so it could be put into a small mold to form an ingot. (Both, author's collection.)

Bethlehem Steel had its own fire department attached to the electrical shop. One plant alone had 71 regular and 1,000 auxiliary firemen. The auxiliary firemen worked their regular jobs throughout the plant. They would inspect, maintain, and educate and train themselves on equipment such as fire extinguishers in their work area. They were successful in putting out 95 percent of all plant fires. The truck is a 1916 hose car made by American LaFrance. In the passenger seat is fire chief Albert E. "Bertie" Anderson, who later left Bethlehem Steel's fire department to be chief at the Bethlehem Fire Department. (Courtesy of Steelworkers Archives.)

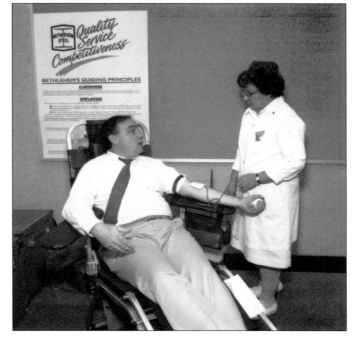

Due to the number of employees on the Bethlehem Steel site, there were multiple dispensaries and medical centers at the main site as well as in Martin Tower and Homer Labs. In 1958, an operation room was added to one of them. Blood donors were always in demand. Here, Bill Fitzpatrick from the sales department gives blood. (Courtesy of the Delaware & Lehigh National Heritage Corridor.)

The No. 2 furnace car makes its way to the blast furnace from the ore yard. It ran back and forth along the Hoover-Mason Trestle multiple times a day since it was an efficient way of transporting material to the blast furnace throughout the Bethlehem Steel site. The No. 2 car remains on the site in its original track location (see page 126). (Courtesy of Steelworkers Archives.)

The recently loaded No. 10 furnace car is getting ready to head for the "pockets" at the blast furnaces. Sinter was the main burden of the furnaces until the introduction of flux pellets. (Courtesy of Ron Keschl.)

A car tip gang is breaking up frozen iron ore on one of its gratings. Material then goes to the ore yard or sinter plant. Sometimes chisels were used, but most of the time, jackhammers got the job done. This job was done mostly in winter. (Courtesy of Steelworkers Archives.)

In extreme cold weather, heaters were placed under the railcars to help thaw the ore and prevent it from freezing and needing to be broken apart. This area was known as the "Thaw Yard" in the 350 Department. (Courtesy of Ron Keschl.)

Once the frozen ore was chipped into smaller pieces, it was dumped. Sometimes the partially frozen load would get stuck in the rail car after the tip. Eventually, it would make its way to the ore yard. (Courtesy of Ron Keschl.)

Once the chunks of ore were dumped from the side dump transfer car in the B pit, a worker needed to get the larger pieces through the grating before the ore went to the blast furnace. These cars were designed to service the car tip and the A and B pits. (Courtesy of Steelworkers Archives.)

A distribution belt for carrying raw materials is seen here at the top of the sintering plant. The belt looks relatively new and clean. In the background is the tripper belt, which sweeps material off the belt into the desired bin. To the left is a tripper operator waiting for ore. (Courtesy of Steelworkers Archives.)

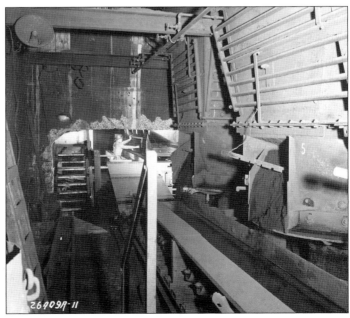

The B pit operator is sending material for the sinter plant. These pit bins are surge bins fed by the car tip or side dump. (Courtesy of Steelworkers Archives.)

This overview of the sintering plant shows the many belts it took to get product moving throughout the plant. The fan house was designed to pull air down the furnace for the combustion unit. The tracks are part of the Hoover-Mason Trestle (wide gauge at 7 feet, 10 inches), which used 250-volt trolley cars. (Courtesy of Steelworkers Archives.)

The sintering plant took in iron ore, limestone, coal, and coke fines. These were then fired to form a cake called sinter, which was charged into the blast furnaces. (Courtesy of Steelworkers Archives.)

Seen here is a Crane check valve in the blast furnace blowing engine house. Although Crane is an American plumbing supply company, the swastika seen here, which appeared on several valves throughout the engine house, means this valve was manufactured in Nazi Germany during the 1930s. (Courtesy of Steelworkers Archives.)

At the top of one sintering building, raw material is being transported on many belts to bring it to its destination. This was transfer belt No. 2, which was a fairly new addition to plant operations. (Courtesy of Steelworkers Archives.)

Steel is poured into an ingot mold to make an ingot. Molds were made of blast furnace cast iron with a refractory lined hot top. Hot steel was poured into them and cooled, and the ingot stripped from the mold was sent hot to the forge or rolling mills. Workers wore protective clothing to protect them from the heat as well as splashes of hot steel. (Courtesy of the Delaware & Lehigh National Heritage Corridor.)

In this 1921 photograph, ingot molds are being stripped from poured ingots under a stripper crane. The molds go to the mold yard and the ingots go to the soaking pits prior to rolling. (Courtesy of the Delaware & Lehigh National Heritage Corridor.)

The third-floor sintering plant shows the Nos. 3 and 4 furnaces (above) moving product to be heated to start the iron making process. The over-fire furnace then ignites and heats a traveling bed mixture of ore, limestone, and coal. It then forms into a cake of iron-rich molecules known as sinter, which is charged in the blast furnace. (Above, courtesy of Ron Keschl; below, courtesy of Steelworkers Archives.)

Two departments work together to supply materials to the sintering plant. The worker on the left is the shuttle for the ore handling department and the worker on the right is the tripper for the sinter plant. This was an early model of control system before it was updated to keep up with the demand for war products. Here, materials are blended together on one belt before it makes its way to the next step in the sintering process. (Courtesy of Steelworkers Archives.)

The iron ore storage yard is spanned by three gantry cranes known as ore bridges with clam buckets to move the ore and limestone. At the bottom of the photograph, railroad cars are lined up and in place to go into forge No. 2. At the top is the Philadelphia, Bethlehem & New England (PB&NE) Railroad locomotive repair shop. (Courtesy of Steelworkers Archives.)

In the Fabricated Steel Construction Division bridge shop, workers assemble (or fabricate) the segments of steel in-house before it goes to the job site to be put into place. They were known as ironworkers, not steelworkers, since they were in a different union. (Courtesy of Steelworkers Archives.)

Although Bethlehem Steel Company made a lot of steel, it also made a lot of the equipment on-site to complete a specific job or task. In the No. 2 machine shop annex, a steelworker is checking out one of the small indoor hydraulic presses that were built on the assembly/test floor. Many of these types of presses and machines went to a materials testing facility, such as Homer Labs, to test the steel being produced. (Courtesy of Steelworkers Archives.)

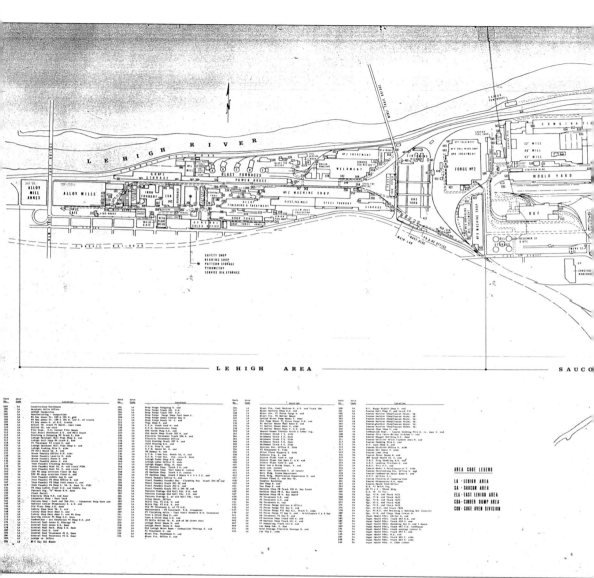

This 1979 map shows the South Side and Saucon Plant of the Bethlehem Steel Company. It was designed by C. Gallo as a guide to help delivery drivers get to a specific loading dock within the

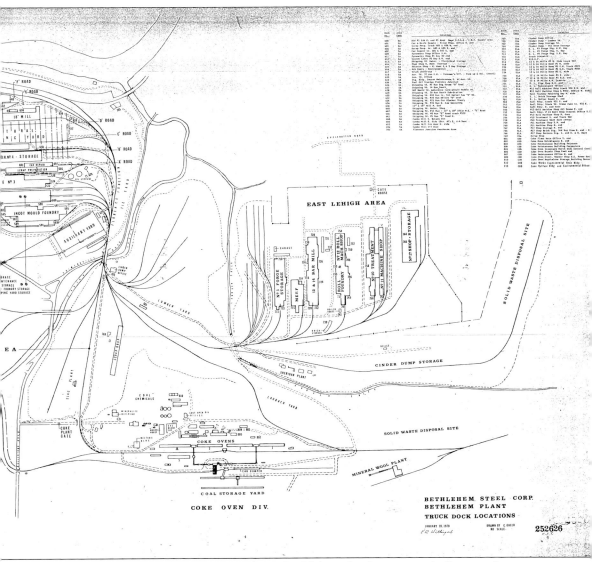

plant. Navigating through a company as big as Bethlehem Steel was quite challenging for visitors as well as employees. (Courtesy of the National Museum of Industrial History.)

The early morning sun shines down on the A battery of the cokeworks above. At right center is the pusher machine, which pushes the coke out of the oven into a waiting car on the other side of the battery. From the A battery quench station, a quench car (below) is receiving a load of coke to be quenched. (Both, courtesy of Ron Keschl.)

An A battery quencher car with a load of coke is being pushed and heading to water at the quenching station. Due to its extreme temperature, especially in the hot summer months, some workers described this area as "hell on earth." Conditions here were almost unbearable, but the employees working there were paid rather well. (Courtesy of Ron Keschl.)

Two doors are being worked on in the A battery door repair shop. Most departments in the plant had "millwright shanties" to repair things when needed. Like many other areas in the plant, this was a dusty and dirty place to work. Nevertheless, some employees took small breaks here. (Courtesy of Ron Keschl.)

Above, the No. 54 ladle car is being filled with hot molten iron from the blast furnace and will make its way to the basic oxygen furnace (BOF). Sometimes gas from the blast furnace built up and caused an explosion, as seen at the E furnace in 1955 at left. As with any other accident, this one was thoroughly investigated so that it could be determined what happened and how to prevent it from happening again. (Both, courtesy of Ron Keschl.)

Schematics, or diagrams, were used throughout the Bethlehem Steel plant to visually see where utilities such as power (pictured), water, or gas were distributed throughout the plant. (Courtesy of Delaware & Lehigh National Heritage Corridor.)

Raw and processed iron ore and limestone were taken to the ore yard. Here, large cranes scooped up the ore, putting it on the internal plant railcars that transported it to the blast furnaces. The ore yard has now been filled in, and the Wind Creek Casino sits in its place. The buildings in the background have been torn down to make room for the facility. (Photograph by Ron Keschl.)

The Olsen machine was used to test the strength of the steel before it made its way through to the next process. This 1923 machine, although simple and easy to use, was replaced over the years with more advanced, computerized machines. (Courtesy of Delaware & Lehigh National Heritage Corridor.)

A crane moves along the ceiling to replace the crankshaft in the 48-inch-wide flange structural mill. The crankshaft directly drives the main mill while the geared shaft in the foreground drives the supplementary (edger) mill. It was designed by the William Todd Co. of Youngstown, Ohio, which built many mill drive engines for Bethlehem Steel Company. (Courtesy of Steelworkers Archives.)

Positioning machines are in place on the assembly No. 1 test floor in the No. 2 machine shop. These machines will be shipped to other steel plants. These pipe machines make a shape and form the steel before it goes on to the next process within the plant. (Both, courtesy of Steelworkers Archives.)

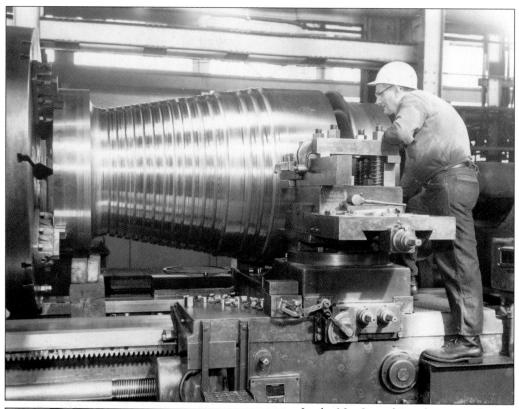

In the No. 2 machine shop, a worker machines a step taper on a forging in a lathe. The speed of the machine is controlled to produce a clean end product with a smooth surface finish. (Courtesy of Steelworkers Archives.)

A large press column in the No. 2 machine shop annex gets nuts bolted on before it goes on a railroad car to its destination in Philadelphia in 1957. Four columns are required to hold the press together so it remains stable. (Courtesy of Steelworkers Archives.)

In the No. 2 machine shop annex, a large pressure vessel is being assembled on the testing floor. The pieces are complex due to many parts being subjected to high hydraulic pressure. (Courtesy of Steelworkers Archives.)

Inspectors and finishers work on a large hollow forging. These jobs usually took place in the Nos. 2 or 8 machine shops due to their size. Each worker had a specific job on every piece to ensure that nothing was overlooked and all was done correctly. (Courtesy of Steelworkers Archives.)

In the No. 2 machine shop, two Carlton radial drills are being used to drill small holes in a sinter pallet for the sinter plant. (Courtesy of Steelworkers Archives.)

A steelworker machines a quarter-circle segment on a large vertical boring mill. These projects took place in the Nos. 2 and 8 machine shops on a regular basis. (Courtesy of Steelworkers Archives.)

In the No. 8 machine shop, a large metal magnet frame is being assembled from several large forgings. They have been machined and are now fitted to be shipped to a US government laboratory, where they will become the core of a Cyclotron atom smasher. (Courtesy of Steelworkers Archives.)

In the No. 8 machine shop, a large gear tire is being machined on a very large vertical boring mill. Most likely, it was shipped to a customer for mounting on a wheel. (Courtesy of Steelworkers Archives.)

In the No. 8 machine shop, a large Ingersoll milling machine finishes a flat forging. (Courtesy of Steelworkers Archives.)

In the No. 8 machine shop, employees are looking at prints of a large assembly that was erected to test fit the pieces before it was shipped to the customer. Bethlehem Steel Company took pride in its quality of work. An inspection department made sure that only quality product was shipped out. (Courtesy of Steelworkers Archives.)

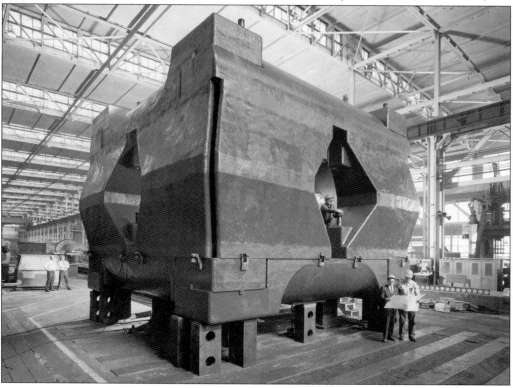

The electric motor repair (AE) shop in 1938 had two overhead cranes and many motor armatures in the foreground. This building has now been restored to be the National Museum of Industrial History, a Smithsonian affiliate showcasing the local steel and textile industries. (Courtesy of Steelworkers Archives.)

In the electric motor repair shop, a Mr. Zambo (left) tests the motor field. These pendant and control stations were used to operate cranes. (Courtesy of Steelworkers Archives.)

Above, AE shop employees pose to celebrate one year without an accident. They surround a four-pole stator, which is part of an electric generator before the armature is installed. The electrical shop team poses again below for another one-year safety award. (Both, courtesy of Steelworkers Archives.)

On a sunny August day in 1941, Adm. C.H. Woodward speaks before steelworkers at the Bethlehem Steel plant before America's involvement in World War II. He was visiting the plant to see how the Navy contracts were being filled. He later visited various areas of the plant to see product as it was being made and talk to those responsible for making them. (Author's collection.)

Executive and office personnel gather in front of a large sign promoting the Liberty Loan Drive. War bonds were sold to help finance the Allied cause during World War I; over $21 billion was raised. Over 100,000 bonds were issued every month. It became a symbol of patriotic duty, similar to jury duty today. (Courtesy of Scott Gordon.)

A finished forging in a heat indication lathe is seen here. The drum is a furnace mounted around the forging in a large lathe. The test piece forging is rotated at a specific speed and temperature to determine its runout, or deviation from design parameters and alignment. This test is required for customer acceptance for critical applications, such as a forging for a steam turbine rotor. (Courtesy of Steelworkers Archives.)

The ingot stripper building housed a steam engine during the pre–World War II era. Here, a crane is taking the mold off the ingots. After this was done, it was then taken to the soaking pits, where the steel was heated throughout so it could easily be formed in a rolling mill. (Courtesy of the Delaware & Lehigh National Heritage Corridor.)

Ernest Bohar is at the controls of the No. 25 electric furnace control shanty. The electric furnace melt shop (EFM) waits for a furnace-charging machine to come into place as steelworkers stand by to work on it. The EFM made many of the largest vacuum-poured ingots ever produced in the western hemisphere. (Courtesy of Steelworkers Archives.)

The No. 3 forge specialty shop shows hot newly forged 16-inch shell blanks that are being quenched in water or oil. This was part of the heat treating for steel to achieve the properties that the customer requested. (Courtesy of Steelworkers Archives.)

An iron foundry worker is pouring a test of iron before molten iron is poured into the row of molds behind him in this 1921 photograph. The iron foundry, built in 1873, used John Fritz's building to make many items for the plant as well as for customers. The building still stands today. (Courtesy of Delaware & Lehigh National Heritage Corridor.)

Product is being loaded from a shipping bay into railcars, to be transported to job sites. The Bethlehem Steel cars were the interchange cars used by Bethlehem Steel and handled by the PB&NE railroad system to deliver to common carrier railroads. (Courtesy of Steelworkers Archives.)

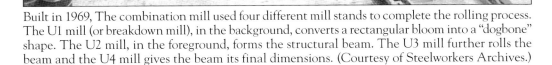

Built in 1969, The combination mill used four different mill stands to complete the rolling process. The U1 mill (or breakdown mill), in the background, converts a rectangular bloom into a "dogbone" shape. The U2 mill, in the foreground, forms the structural beam. The U3 mill further rolls the beam and the U4 mill gives the beam its final dimensions. (Courtesy of Steelworkers Archives.)

While the steel is still red hot, a hot saw cuts through a heavy column section after it is finished rolling from the 48-inch structural mill. The hot (and cold) saws were over six feet in diameter. This job was done by the beam yard workers. (Courtesy of Steelworkers Archives.)

The image above of the No. 8 double acting, 6,000-pound steam hammer shows the four column, double frame hammer and a manipulator on each side to hold the work piece. Below, workers in the tool steel department melt shop pose for the camera. (Both, courtesy of Steelworkers Archives.)

These iron foundry workers are using air-powered rammers and are molding the sand by packing and compacting it into a flask. It surrounds a wood pattern that was made by the pattern makers in the pattern shop. (Courtesy of Steelworkers Archives.)

A welfare room was in every shop at the Bethlehem Steel site. It included bathrooms, showers, and changing areas so that employees could wash and change before going home to their families. Baskets were drawn up to the ceiling by chains, sometimes secured with a lock. Employees usually had two baskets: one for work clothes and the other for street clothes. (Courtesy of Steelworkers Archives.)

Workers make their way out of the plant and punch a time clock after the company whistle goes off in this 1949 photograph. Plant patrol personnel (who were sometimes women) were at all the gates to ensure the safety and security of workers and the facility. They also checked bags for theft of goods. (Author's collection.)

From left to right in front in this Ed Leskin photograph, steelworkers Pete, Bernie, Barry, Dave, and Bob make their way out the main gate at the Fillmore Street entrance to the Bethlehem Steel plant. (Courtesy of Steelworkers Archives.)

Three
OFF THE CLOCK

In any steel company, clambakes and picnics were popular events at the local park pavilions. Most times, they were paid for by the company or union and included endless food and beverages. (Courtesy of Steelworkers Archives.)

The grandstand and bleachers of the Bethlehem Steel company sit a few miles from the Bethlehem plant on Elizabeth Avenue off Center Street. Many athletic events took place there. The facility was given to Moravian College (now Moravian University) in 1962 and is now called Rocco Calvo Field and Timothy Breidegam Track. (Courtesy of Scott Gordon.)

Members of the Steel F.C. soccer team pose after a win at Taylor Field near Lehigh University in 1915. The team started as a recreation club for Bethlehem Steel employees in the early 1900s. They had many recruits from Scotland and England, and won eight championships, six american cups, and five national challenge cups. Bethlehem Steel sponsored the team until 1930, when the company vice president went to another firm. (Courtesy of Daniel Paul Morrison.)

The Bethlehem Steel softball team was formed to provide family entertainment during World War I. Members of the team included workers from the plant mixed in with a few pros and minor league players. Games were every Saturday from May through Labor Day. (Courtesy of LehighValleyLive.com.)

Before social media, workers had to rely on radio to get the news. Louis Horvath and George Blair were on unpaid picket duty and took time to tune in during the 1949 strike. (Author's collection.)

The Bethlehem Steel Company band hall was where band members would meet and practice. It was then the Bethlehem Area Public Library before it moved to its current Church Street location. (Courtesy of Scott Gordon.)

The Bethlehem Steel band poses for a group photograph in the early 1900s. The band performed in parades, concerts, and various events throughout the year. It was made up of steelworkers from throughout the plant and offices. (Courtesy of Scott Gordon.)

The band poses before performing for the superintendent's clambake of 1911. These events usually took place at a local park pavilion and provided hours of fellowship and entertainment among the steelworkers. (Author's collection.)

As the years went on, the Bethlehem Steel band got more creative and decided to use a miniature horse to pull the drum in parades instead of having the drummer carry it. They performed in many parades and made special appearances at the plant site. (Author's collection.)

The Bethlehem Steel Country Club was started in 1934 by a group of supervisors. In 1946, adjacent property was acquired for a golf course. It was a common meeting ground for executives and a place to take clients for a round of golf or dinner. In 1986, it became Silver Creek Country Club, and in 2018, it became the Steel Club. (Courtesy of Scott Gordon.)

Saucon Valley Country Club opened in 1920 with a group of 16 business leaders, mostly from Bethlehem Steel, who bought 270 acres along Saucon Creek. It was a similar place for upper management and executives. It also had black Angus cattle on the grounds to provide steaks for members. (Author's collection.)

In 1955, the Fellowship Club of Bethlehem Steel Company performed its annual play during its sixth annual fellowship minstrel. The cast and crew were made up of current steelworkers in the plant and offices. (Courtesy of the Delaware & Lehigh National Heritage Corridor.)

In 1957, the Fellowship Club performed "The Green Door" at the Bethlehem Steel Club during its eighth annual fellowship minstrel. It included an orchestra and a quartet under the direction of F. Mies and technical director P. Remely. (Courtesy of the Delaware & Lehigh National Heritage Corridor.)

Two twin-engine planes owned by Bethlehem Steel Company were used to take executives to meetings or vacations as well as to deliver parts in a hurry. They were based at the Allentown, Bethlehem & Easton Airport (presently Lehigh Valley International Airport) and would log over 150,000 miles in a year. (Courtesy of the Delaware & Lehigh National Heritage Corridor.)

The company hired the best pilots and mechanics available. The planes' interiors were nothing short of the best. (Courtesy of the Hagley Museum,)

The Bethlehem Steel private planes wait near the hangar. Airport Road is in the background, in the vicinity of today's Day's Inn. Over time, Airport Road was rerouted and homes were torn down to make larger runways. (Courtesy of the Delaware & Lehigh National Heritage Corridor.)

Gray Mill workers prepare a float to help promote liberty loan bonds. Liberty bonds were issued through the Department of Treasury and Federal Reserve in 1917–1918 to help finance World War I and the Allied effort in Europe. They came back into existence after September 9, 2001, to help rebuild Ground Zero as well as other locations that were affected. (Author's collection.)

Steelworkers celebrate and burn picket signs at the temporary tent that was set up at the main gate. The CIO agreement announcing the end of the strike came on October 31, 1949. Picketers were off the time clock and unpaid. They were elated that an agreement was made and they could return to work. (Author's collection.)

Plant workers pose for a group photograph after the end of their long shift. Uniforms were not given to employees during this time. The dress code was whatever the employee could afford or what they were comfortable in. Protective clothing was provided by the company for jobs requiring it. (Courtesy of Frank Behum Sr.)

Four
THE PEOPLE

Henry Altemose (left) and Walter Deal are at the controls of the blast furnace. It was their job to ensure that the blast furnace ran smoothly and without issue. They monitored the heat and scheduled times when ovens were to be pushed. There was also a small area here used as a break room for the department workers. (Courtesy of Ron Keschl.)

Hanging out in the break room at the blast furnace are Joe Cerquiera (left) and Rich Roth. They were taking a break between casts, which could be approximately 90 minutes. This was the day of the last cast, November 18, 1995. (Courtesy of Ron Keschl.)

Workers are at the C blast furnace preparing the last cast of hot molten iron ever to take place at the Bethlehem Steel plant in 1995. Malik Nadraws (left) and John Cressman monitor the blast furnace. Special protective clothing was worn in this department to prevent molten metal splash injuries and keep employees protected while working in extreme temperatures. (Courtesy of Ron Keschl.)

Bob Holly was a joint safety and health inspector at the basic oxygen furnace, a charging crane man, and hot metal teeming person at Bethlehem Steel Company from 1973 to 1995. He was well known throughout the company and put in 22 years of service. (Courtesy of Bruce Ward.)

Employees of the No. 2 machine shop pose for the camera in this 1930s photograph. Much of the labor at this time was manual, but large cranes were used to move very heavy items throughout the facility. The inset photograph shows the exterior of the machine shop. (Courtesy of Frank Behum Sr.)

Steve Kuscon is the only one known who videotaped the final operations of the last cast in the blast furnace in 1995. He was a stove tender on the night shift and remained at the plant that day for the last cast. His wife, Donna, also worked on the cast floor until her layoff. (Courtesy of Ron Keschl.)

Lenny Neith (left) leans over the table to shake hands with Eddie Mauer on the last day before his retirement as his coworkers (from left to right)C. Fritz, R. Keschl, C. Newhart, and N. Angst look on. Mauer was an electrical engineer for the sinter plant, ore yard, and car tip. The break room was not just used for lunch, but also for meetings and updates. Most buildings at the plant had a break room. (Courtesy of Ron Keschl.)

Walter Hotltzschafer is tying up the blast furnace bell with a chain to go in and work on the furnace. The job of a rigger was dangerous, since toxic gases were being vented while working on the blast furnaces, but no two days were exactly the same. (Courtesy of Bruce Ward.)

The Lewis girls, with the help of their mother, relight the blast furnace. Blast furnaces rarely went down due to the production schedule. It was a big deal when one was relit. Usually, children of top executives would be given the honor of lighting them. (Courtesy of the Delaware & Lehigh National Heritage Corridor.)

Prominent Manhattan lawyer Paul Drennan Cravath, from Blantchford, Seward & Griswold, is with Charles Schwab and Eugene Grace outside a courthouse. One of Cravath's earliest cases concerned Thomas Edison's incandescent light patent at Edison Illuminating Company. Other clients included the Baltimore & Ohio Railroad, Studebaker Corporation, and of course, Bethlehem Steel. In 2019, his law firm celebrated its 200th anniversary. (Author's collection.)

President of the US Steel Corporation and president and chairman of Bethlehem Steel Charles M. Schwab is being filmed with his wife, Emma Eurana Dinkey, in his 1927 Packard Phaeton. Schwab was always known to attract a crowd when he and his wife were in the area. (Courtesy of Nick Chismar.)

Edmund F. Martin (left) was the chairman and chief executive officer of Bethlehem Steel Company from 1964 to 1970. He joined the company as a member of the Loop Management Program in 1922. The Martin Tower office building in North Bethlehem was named in his honor. Arthur Homer (right) was chief executive officer from 1957 to 1964, during which time he broadened the corporation's research work. Homer Labs was named for him. (Author's collection.)

Donald Trautlein started his career with Bethlehem Steel in 1977. A short two years later, he was elected chairman and became chief executive officer in 1980. He is joined here by Lynn Williams, president of the United Steelworkers International Union, during a press conference in Pittsburgh responding to the decision that the US International Trade Commission made restricting steel imports for five years. (Author's collection.)

John A. Stephens, vice president of US Steel; Phillip Murray, president of CIO and United Steelworkers; and Joseph M. Larkin, vice president of Bethlehem Steel Corporation hold a press conference with the United Steelworkers as they negotiated to try to end a 49-day strike. Murray joked to the media during the press conference, saying, "Murray in the Middle." (Author's collection.)

From left to right, Kazimier Miller, Howard Curtis, and Lawrence Schaffer look over plans calling for an immediate strike, with Brendon Sexton, Frank Fernback, Erwin Leppert, and Hubert Atallah looking on in the second row. The CIO alleged that Bethlehem Steel had a company union and was preparing for the election of officers. (Author's collection.)

Archibald Johnston was a mechanical engineer who worked his way up the ladder at Bethlehem Steel and became president. He was also the first mayor of a united Bethlehem and was responsible for building the Hill to Hill Bridge. His wife, Estelle Sophia Stadiger Borhek, was active in the Central Moravian Church. His daughter Elizabeth Johnston, at the age of 17, worked at Bethlehem Steel to help out during the war with no pay (her salary was donated to the American Red Cross). Archibald Borhek Johnston ("Arch Jr.") became an avid photographer. His mansion still sits on 500 acres in Bethlehem Township, known as Camels Hump Farm. It is currently being restored. The land is being conserved through the township called the Archie Project and offers trails for hiking, birdwatching, and other recreational activities. (Courtesy of John Marquette.)

A celebration of 40 years of service with a dinner reception was one of many special events to take place at Bethlehem Steel. Mike Zaia (left) worked for Bethlehem Steel for over 40 years, finishing in Martin Tower, and was one of the last ones to leave the building. He is joined by Tim Lewis (center) and Don Trexler. Seated are Zaia's wife, Maria Zaia (left), and Pat Trexler. (Courtesy of Steelworkers Archives.)

Inspector of the Argentine government, Charles Lynn (left), visits the Bethlehem Steel No. 2 machine shop assembly floor. The foreman (center) took Lynn and Carl Allgum throughout the facility to show them how steel is made. (Courtesy of the Delaware & Lehigh National Heritage Corridor.)

A plant employee is demonstrating armature repair in the electrical shop to a group of "loopers" from Lehigh University as a supervisor looks on. Loopers were newly hired engineering graduates who "looped" through the plant to learn all operations, machines, and duties. (Courtesy of Steelworkers Archives.)

Escorts pictured on the marble steps at the Steel General Offices building (above) were modeled after flight attendants and served as ambassadors, greeters, and elevator operators and to escort visitors to their destinations in the office buildings. They were trained by Barbizon Modeling School on poise, personality, and self-care. Their dresses were of the best quality and material from New York City boutiques. They all had to be a certain height, and even a certain size (12). Once they started to show signs of pregnancy, they lost their job. Below, at the Saucon Valley Country Club, escorts work at a Christmas party, serving as greeters with a warm, welcoming smile as guests enter the ballroom. (Both, courtesy of Fran P. Bush.)

During World War II, there was a high demand for security guards. Bethlehem Steel had many government orders to fulfill and needed extra security to keep things secure throughout the plant. These women were armed security guards, and were paid well. A few even shot better than their male colleagues. They were nationally known as "the Pistol Packin' Mammas." (Courtesy of Steelworkers Archives.)

Gite Christian moved to Bethlehem from the small town of Quitman. He shortly received an opportunity at Bethlehem Steel as a plant patrol officer. He worked at various plant facilities, including Burns Harbor. The officers were armed, but fortunately, in his 18 years of service, Christian never had to take out his gun. (Courtesy of Ron Keschl.)

While most female employees worked in the office and most males worked in the plant, there were some exceptions. Robert Seifert (right) was one of the first male receptionists at Bethlehem Steel in Martin Tower, pictured with Fran P. Busch, secretary at Martin Tower. Pat Dwinell (below) worked in the blast furnace as a repairman and electrician. She met her husband, George, while working in the blast furnace division. (Right, courtesy of Ron Keschl; below, courtesy of Fran P. Bush.)

The shipping yard crew take time from one of their breaks to pose for the camera; from left to right are Fritz Zondlo, Steve Michaleryh, Walter Nizio, John Toth, Al Nemeth (standing), Lew Mongilutz, and Bill Magdasy. Many workers created strong bonds due to the long hours they worked together. (Courtesy of Steelworkers Archives.)

Electrician Donald Serfass crimps a connector on the end of a wire in front of a control panel for his department's intercom system. It was the most efficient way to communicate with employees throughout the department. Plant-wide communications required the Bell Telephone System. (Courtesy of Steelworkers Archives.)

Bethlehem Steel's electrical school took place on the second floor in an old machine shop. It was a year of instruction followed by on-the-job training or an apprenticeship, gaining the hands-on experience needed to work on the electrical systems at the plant and in the office buildings. Martin Schafer was a motor inspector in the 415 Department before he became the electrical school instructor. (Courtesy of Bethlehem Area Public Library.)

The welding school trained thousands of men and women in the essential skills needed to be a valuable asset to the plant with knowledge of the various types of welding techniques. A group of welders in training in the early 1990s pose outside one of the office trailers for a quick photograph before class. (Courtesy of Bruce Ward.)

Rigger Bruce Ward fixes some welds on one of the crane rails high above the ore yard. Some of the qualifications to become a rigger included the ability to climb, not have a fear of heights, and the ability to work outside in all types of weather. Ward spent 30 years working at Bethlehem Steel until it closed. (Courtesy of Bruce Ward.)

Bethlehem Steel Company workers pose outside the No. 1 thirty-two-inch mill and soaking pits. It was the last week they were on the job at the company. Workers in this department were only allowed to work for short periods of time and were given mandatory breaks due to the extreme heat produced in this building. (Courtesy of Steelworkers Archives.)

Steve Cengeri's coworkers pose for a photograph in the No. 2 machine shop west-end layout plate department. From left to right are (first row) Otis Walhoffer and Richard Buss; (second row) Jim Saeger, Roger Bodnar, Gordan Haney, Bill Cengeri, and Bill Shafnisky; (third row) Milton Arnold and Luther Hottle; (fourth row) Harry Graaf. (Courtesy of Steelworkers Archives.)

Richard "Richie" Check spent 43 years at Bethlehem Steel as a rigger. He was very passionate about talking to people about the history of Bethlehem Steel and his experiences there. After his retirement, he was very active with the Steelworkers Archives as a historian and tour host. He was interviewed on many occasions. (Courtesy of Steelworkers Archives.)

John Fandl Sr. (left) was a carpenter at Bethlehem Steel's Saucon Division for 15 years. His son Louis Fandl worked in the sintering plant for 15 years as well. His grandson Ron Keschl (below) got his first job fresh out of the US Marines at Bethlehem Steel, where he worked for 34 years in many departments, including the sintering plant, ore yard, environmental department, blast furnace, and coke works. Having multiple generations as well as entire families working at Bethlehem Steel was very common. (Both, courtesy of Ron Keschl.)

Robert Sayre helped found the Bethlehem Iron Company, where he was the general manager. He gave generously to Lehigh University, Sayre Observatory, and St. Luke's Hospital. He built a large house in Bethlehem in 1858 where he lived until his death in 1897. It is presently a bed and breakfast and event facility near the Hill to Hill Bridge. (Courtesy of Sayre Mansion.)

This group photograph of the United Steelworkers union was taken at the 50th convention anniversary. The United Steelworkers of America (USW) was founded in 1942. A few months later, three different unions were created at the Bethlehem Steel plant due to the 31,000 employees working there. Local Nos. 2598, 2599, and 2600 were created on May 22, 1942. (Courtesy of Steelworkers Archives.)

Bartholomew "Bart" Check, also known as "Checkie" to his steel-working colleagues, poses outside the Bethlehem Steel plant's parking lot next to a company truck. He, as well as several brothers, worked at the plant, including Emil in the carpenter shop and Richie, who was a rigger (see page 79). (Courtesy of Bruce Ward.)

The management team poses on the steps of the Bethlehem Steel Country Club in Hellertown, Pennsylvania, in September 1971. From left to right are (first row) William "Bill" J. Ziegenfus, turn foreman at Saucon Mills; Owen H. "Os" McClafferty, turn foreman at Saucon Mills; Bernard A. Duddy, chief clerk at Saucon Mills; Anthony T. "Tony" Raczenbek, turn foreman at Saucon Mills; Peter A. "Pete" Zelinsky, industrial analyst at industrial engineering; and J. Richard "Dick" Langkamer, chief planner at production scheduling; (second row) Louis "Lou" Kercsmar, turn foreman in manufacturing; William "Smokey" Sandrovitz, turn foreman in manufacturing; William J. "Bill" Bernhard, turn foreman at Saucon Steelmaking; Martin J. "Marty" Jandris, turn foreman at Saucon Steelmaking; and William "Bill" Kuzmin, foreman at manufacturing; (third row) Amos L. Proctor, instructor; William B. "Bill" Turner, safety engineer at industrial relations; William C. "Bill" Siegfried, safety engineer at industrial relations, Albert "Al" Wachinski, turn foreman at Saucon Mills; and John B. "Jack" Stevens, instructor in manufacturing. (Courtesy of Chris Rosati.)

Ron Keschl (left) poses with some of his coworkers on the D furnace during one of the last days that casting took place at the Bethlehem Steel plant in 1995. (Courtesy of Ron Keschl.)

David Blackwell, Bethlehem Steel plant general manager in the 1980s, looks over whistles with a few of his colleagues. A British-made whistle was recovered and used at the Bethlehem plant when there was a shift change. In 1952, it sounded for two hours straight and was removed. It was sent to Brooklyn Pratt Institute where, on special occasions, students and staff would set it off. The whistle is now on display at the Maritime Industrial Museum at Fort Schuyler. (Courtesy of Steelworkers Archives.)

Workers gather for a group photograph under an overhead crane. (Courtesy of Frank Behum Sr.)

Bruce Davis joined Bethlehem Steel in 1964 as an attorney in the sales department. He was involved with the expansion of Route 33 from Route 22 to Route 80 in 1969. Although he left the company in 1985 to work at a local law firm, he continued to support the expansion of Route 33 as well as the long-term planning to reach Interstate 78 in 2002. (Courtesy of the Easton Area Public Library Archives.)

Five

PRODUCTS, PROJECTS, AND TOOLS

One of the largest forged products Bethlehem Iron ever made was a Ferris wheel axle in 1893. It took two years to complete, weighed 86,320 pounds, and was 33 inches in diameter and 45.5 feet long. It was built for the 1893 Chicago world's fair where, over 19 weeks, a record 1.4 million people rode it for 50¢ per passenger. (Author's collection.)

San Francisco's mile-long Golden Gate Bridge is one of the most famous landmarks in the world, but it was also one of the most challenging to build. Weather, strong tides, and the nearby San Andreas Fault were just some of the challenges. McClintic Marshall was the fabricating and steel bridge–making company that was acquired by Bethlehem Steel in 1931. Over 45,000 tons of steel were used for this project, which was shipped through the Panama Canal. (Author's collection.)

Here, finishing touches are being done to the bridge. It was completed in 1937, with a day of foot traffic followed by automobile traffic the next day. It was painted orange to help with visibility in the fog and other severe weather. (Author's collection.)

Spanning the Hudson River between New York and New Jersey, the George Washington Bridge was designed by Othmar Ammann, a Swiss-born architect and engineer. Design started in 1923, and construction started in 1927. The 3,500-foot center span is suspended between two 570-foot Bethlehem Steel–fabricated towers. (Author's collection.)

The opening-day celebration of the George Washington Bridge in 1931 included festivities before the first car went over. Over 5.5 million vehicles drove over the bridge in its first year. Fifteen years later, the two center lanes were paved, and finally, in 1962, all six lanes were finally opened for better capacity. (Author's collection.)

The Rockefeller Center's 19 buildings used Bethlehem Steel throughout the 22 acres of property. The iconic photograph *Lunch atop a Skyscraper*, with 11 steelworkers sitting on a Bethlehem Steel beam having lunch, was taken here on September 20, 1932, on the 69th floor. Rockefeller Center took over nine years to build. (Author's collection.)

Immediately after Harry S. Truman's election, the first family, along with the White House staff, moved across the street to the Blair House. The entire East Room of the White House was gutted down to the Bethlehem Steel support beams as part of Truman's large scale renovation from 1948 to 1952. A souvenir program included nails, bricks, stones, and even burnt wood, which raised over $10,000 by 1951. (Courtesy of Abbie Rowe and the White House Historical Association at Decatur House, a National Trust Historic Site.)

This newspaper ad shows that the Bethlehem Steel Company produced the steel forgings for the Saturn rocket transporter for NASA's Apollo missions. A total of 176 forged-steel rollers weighing over 2,000 pounds each were made at the Bethlehem plant. (Courtesy of Steelworkers Archives.)

The Waldorf Astoria Hotel in New York City was built in 1931 at 301 Park Avenue between 49th and 50th Streets. It took 27,100 tons of steel to build it and 300,000 cubic feet of limestone for the building's facing. It is world-renowned for its luxurious Art Deco design as well as the many dinner parties and galas held here over the years. (Author's collection.)

St. Luke's Hospital on Fountain Hill dates to 1872. At that time, there was a need to care for patients injured at Bethlehem Steel as well as local cement mills, coal mines, and explosives factories. A major addition was built in 1946 to accommodate the influx of workers and their families. This project took a few years to complete. (Courtesy of Bethlehem Public Library Archives.)

The Bethlehem Area Public Library and city hall are being constructed using Bethlehem Steel in 1966. A few months later, in 1967, over 400 volunteers helped move over 80,000 books from the old Market Street facility to the new location. Children from area schools as well as local scouts helped to carry the remaining 20,000 books that were too big to pack in boxes. (Courtesy of Bethlehem Public Library Archives.)

Steelworkers show sections of interlocking PZ piling produced by the combination mill. Used primarily for retaining walls, the fingers on the tops of the piling interlocked with the bulb on the bottom of the adjacent section as they were driven to form a wall. (Courtesy of Steelworkers Archives.)

The French ship SS *Normandie* was a luxurious ocean liner that had a record-setting maiden voyage to New York. With the fear of it being sunk during World War II, it was taken over by the US government in 1941 and renamed the USS *Lafayette*. A year later, it caught fire and capsized. The ship's remains were sent to Bethlehem Steel by railcar to be melted down to make more steel. (Author's collection.)

The USS *Massachusetts* was built at the Bethlehem Steel shipyard in Quincy, Massachusetts. It was launched in 1942 and fought in numerous battles in the Pacific. Also known as "Big Mamie," it was saved from being scrapped. A group of Massachusetts schoolchildren raised money to save it, and two months later, it was opened to the public at Battleship Cove in Fall River, Massachusetts. (Author's collection.)

The USS *Nevada* was launched in 1914 and served in both world wars. Although the ship sank, it was salvaged and modernized. In 1946, the ship was exposed to an atomic bomb blast during a test and was heavily damaged and made radioactive. It was decommissioned shortly after and sunk during gunfire practice two years later. (Author's collection.)

Adm. C.H. Woodward of the US Navy talks to a steelworker in this 1941 photograph. The admiral, along with several aides, toured the plant to see how the Navy contracts were being fulfilled. Bethlehem Steel played a huge part in World Wars I and II. (Author's collection.)

Although it was not made at the Bethlehem plant itself, barbed wire, also known as steel fencing, was produced at other Bethlehem Steel plants throughout the United States. Bethlehem Steel's Johnstown plant made this special wire that was used primarily for animal enclosures. (Courtesy of the Delaware & Lehigh National Heritage Corridor.)

Above, a hollow-shaft forging is making its way from the machine shop to the High House on the plant's internal rail system in the early 1900s. At left, the No. 5 High House was a tall, skinny building that was used to accommodate large items that needed to be heat treated vertically. Large cranes lifted the items from rail cars. (Both, courtesy of Steelworkers Archives.)

A disappearing coastal defense gun is being assembled in the No. 2 machine shop. Many times, the entire No. 2 machine shop was lined with material for war production. Many large gun components were made by Bethlehem Steel. (Courtesy of Steelworkers Archives.)

A steelworker is operating the Landis roll grinding machine in the forge specialty department (No. 3 forge). After machining and hardening, the bearing and working surfaces of a high-quality metal roll were ground to a mirror-like finish. (Courtesy of Steelworkers Archives.)

The USS *Northampton* was a heavy cruiser built in 1929. It was sunk by Japanese torpedoes on November 30, 1942, during the battle of Tassafaronga. It was named after Calvin Coolidge's hometown of Northampton, Massachusetts. (Author's collection.)

The automotive industry benefited a great deal from Bethlehem Steel. Although not produced at the Bethlehem plant, automotive sheet steel was made and shipped to car manufacturing plants throughout the country to be formed and assembled on production lines. (Courtesy of the Delaware & Lehigh National Heritage Corridor.)

Six
HOMER LABS, SGO, BPO, AND MARTIN TOWER

This aerial photograph of the Homer Research Labs shows how large an area it covered. It opened in 1961. At upper left is a small part of the Bethlehem Steel plant. (Courtesy of Steelworkers Archives.)

Employees of the Homer Labs Research Department and allied services personnel pose on a spring day in 1967 in front of the manmade pond adjacent to Iacocca Hall. Today, the hall contains classrooms and a reception area for events offering a panoramic view of the Lehigh Valley. (Courtesy of Steelworkers Archives.)

Building C consisted of high bays C1, C2, and C3, which offered ample overhead room. The property had eight buildings that spanned over 1,000 acres at the top of South Mountain. Today, the buildings are mostly owned by Lehigh University. (Courtesy of Steelworkers Archives.)

Robert Fenstermaker takes a coffee break in his Homer Research Lab office in 1972. He was a plant worker before he transferred to Homer Labs to become a chemist. He retired in 1980 after 42 years of service. (Courtesy of Linda Finken.)

This large group photograph was taken on the stairs of the lobby of the Steel General Office (SGO) building. The group was attending a Bethlehem Steel management meeting on February 13, 1950. This was a popular location for formal group pictures by Bethlehem Steel photographers. (Courtesy of the Delaware & Lehigh National Heritage Corridor.)

In the early 1900s, the Steel General Office building was erected to accommodate the upper management as well as the executives, CEO, and president of the company. Four floors were initially built (above), and then a few years later, in 1916, more were added and the building was extended (below). It included elevators, an auditorium, dining facilities, conference rooms, and many offices. (Both, courtesy of Scott Gordon.)

The SGO, also known as the East building, was erected rather quickly during World War I. It was expanded to 13 stories in an "H" shape in 1916, and expanded again in 1928. The final addition of the lobby and rear annex was built around 1950. (Courtesy of Scott Gordon.)

With the Bethlehem Steel Company growing in the 1950s and 1960s, more floors were added to the top of the SGO building. Currently, the building is still standing but has an unknown future due to its deteriorating condition. (Author's collection.)

In SGO building room 493, secretaries are busy at their desks answering phone calls on their individual switchboards. A large directory was on each desk to quickly find the names and numbers to connect the caller to the required employee or office. When these photographs were taken in 1969, Bethlehem Steel Company had the best technology possible for its office staff. (Both, courtesy of Steelworkers Archives.)

Office workers gather for a photograph with the "tub file," which sorted the IBM cards that were loaded into a computer to process data. These early data cards did not hold much information. (Courtesy of Steelworkers Archives.)

These women at a switchboard are overseeing office operations through closed-circuit TV. They were able to patch callers through to a person or department. The TV monitors were used for quality-assurance monitoring as well as employee safety. (Courtesy of Steelworkers Archives.)

The top of the SGO building offered a panoramic view of most of the Bethlehem plant site. The image above shows the view to the west, and the image below shows the view to the north. This was the tallest building in Bethlehem until Martin Tower and a few apartment buildings were built. (Both, courtesy of Scott Gordon.)

This construction photograph of the West building, also known as the Bethlehem Plant Office (BPO), was taken from the top floor of the SGO building in the 1940s. A few years later, another addition was added to the SGO to keep up with the growing demand for office staff at Bethlehem Steel. (Courtesy of the Delaware & Lehigh National Heritage Corridor.)

The two main office buildings on the Bethlehem Steel site were less than a block from each other. Beside one of the plant entrances was the West building, which was mostly sales offices until Martin Tower was built. It then became a plant office. When Bethlehem Steel closed, it became Northampton Community College's Fowler Center. (Author's collection.)

Fresh Christmas trees were displayed in all office building lobbies, including the SGO building, during the holidays. They were bought in quantity from a local tree farm and set up with the help of the plant's horticultural department. (Courtesy of the Hagley Museum.)

The lobby has marble flooring as well as a marble desk in the center, where the receptionist greeted visitors, clients, and customers. The limestone sculptures on the lobby wall were designed by Sidney Waugh, who worked for Steuben Glass and did pieces for several federal courthouses. They portray suppliers, consumers, employees, management, stockholders, and two divisions of the plant. (Courtesy of the Hagley Museum.)

Fine artwork was placed throughout the various office buildings at Bethlehem Steel. This painting by Dean Cornwell shows steelworkers at a blast furnace. Fresh floral arrangements also added a decorative touch throughout the buildings. (Courtesy of the Hagley Museum.)

A.B. Homer stands on stage before a crowd to dedicate the Grace Auditorium. It was used for large gatherings, meetings, and corporate announcements and events. It had amplifiers, "roving" (wireless) microphones, and duplicate recording machines, as well as 35mm and 16mm motion picture projectors. Architects for the project were McKim, Mead & White of New York. (Courtesy of the National Museum of Industrial History.)

In 1920, a coffee shop was on the lower floor and offered coffee and light snacks to office employees. The silver coffee urns, wooden stools, and fresh plants arranged around the bar added a decorative touch. (Courtesy of the Delaware & Lehigh National Heritage Corridor.)

The snack bar was set up so employees could get a quick meal or a cup of coffee. The food was affordable, and daily specials were offered and announced in advance so a worker could decide whether or not they would bring their lunch from home that day. (Courtesy of the Hagley Museum.)

The cafeteria in the basement served an array of foods for the office staff. Entrees were made from scratch, and quality ingredients were used. Fresh potted plants served as centerpieces on the tables and were replaced regularly. (Courtesy of the Hagley Museum.)

The executive dining room had the finest china, silverware with the Bethlehem Steel logo, and pressed napkins and linens adorning the tables. The waitstaff who worked here serving the executives enjoyed the perk of also eating the finest foods when they took their lunch breaks. (Courtesy of the Hagley Museum.)

There were several mail rooms throughout the Bethlehem Steel site. Pictured here is the SGO building mailroom. There was also one in the plant as well as at Martin Tower on the north side. Mail was sorted and distributed by two different plant couriers: plant mail and production department mail, either via the plant bus or on foot throughout the site. (Courtesy of the Delaware & Lehigh National Heritage Corridor.)

The Eighth Avenue North Office was at the corner of Eighth and Eaton Avenues before Martin Tower was built. The live plants were meticulously maintained by Bethlehem Steel's horticultural department, which had an open account with the local Sawyer and Johnson floral shop. (Courtesy of the Hagley Museum.)

This meeting room includes replicas and artist renderings of the future Homer Labs and other proposed projects that were in the works. The curtains were made of custom printed fabric for the Bethlehem Steel plant. The conference room had ashtrays on the meeting tables so that work could be done without taking a cigarette break. (Courtesy of the Hagley Museum.)

The office of Eugene Grace had the finest artwork, furniture, and décor on display. It had a working fireplace and overlooked the plant through a window. When Grace went to his office, all elevators were emptied out on the closest floor and sent down to the lobby so that he could choose which one to go up in. (Courtesy of the Hagley Museum.)

The Ship Room showcased many of the ships that Bethlehem Steel produced during the wars. For many years, historians and archivists were convinced that these model ships were long gone. They were discovered recently and are now on permanent display at the Massachusetts Maritime Academy near Cape Cod. (Courtesy of the Hagley Museum.)

Richard Beidleman and his son Rick look over the display of ships in the Ship Room. Beidleman worked in the building and took his son to work one day to show him the office. In 1943, there were 380 fighting and cargo ships made by Bethlehem Steel. (Courtesy of the Hagley Museum.)

Martin Tower was the corporate headquarters of Bethlehem Steel. It was named after former chairman Edmund F. Martin and had an occupancy of 1,300 employees. It towered over the Lehigh Valley at 21 stories until it was imploded on May 19, 2019. (Author's collection.)

Martin Tower and its neighboring buildings were built in several phases over the span of a few years. The final amounts of each project are shown here. While it cost over $43 million to build Martin Tower in 1973, it was demolished at a cost of around $500,000 in 2019. (Courtesy of Al Fetterolf.)

Harold S. Campbell sold 53 acres to Bethlehem Steel to build Martin Tower. Ground was broken in 1970, and construction started right away using the iconic Bethlehem Steel wide flange beams. A line of homes along Eighth Avenue were used as offices for the contractors during construction. The homes were all later relocated a few blocks away. The Durkee spice plant in the background closed in 1995 and is now the site of Lowes Home Improvement Center. (Courtesy of the National Museum of Industrial History.)

Fran P. Busch (right) assists a client visiting the Martin Tower reception area. An indoor walkway connected Martin Tower's main lobby to wings A and B and the Stewart S. Court Auditorium. They were commonly referred to as the "Annex buildings" by the steelworkers. (Courtesy of Fran P. Bush.)

Fran P. Busch greets two employees at her reception desk. She was one of the first secretaries to work the main desk at Martin Tower. She joined Bethlehem Steel in 1965 as an escort working in the SGO building on Third Street until Martin Tower was erected. (Courtesy of Fran P. Busch.)

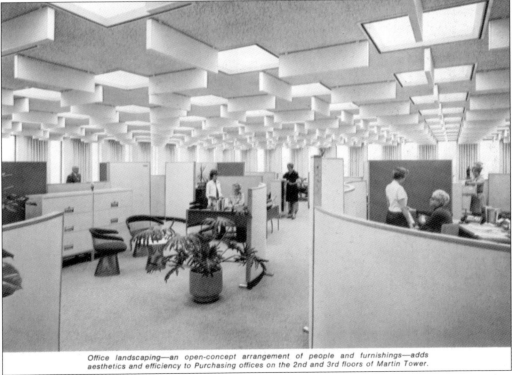

Office landscaping—an open-concept arrangement of people and furnishings—adds aesthetics and efficiency to Purchasing offices on the 2nd and 3rd floors of Martin Tower.

An open concept design was established in the offices of Martin Tower. Partitions were used to make cubicles, and much of the furniture was made with Bethlehem Steel. The unique cruciform building shape allowed more offices to have windows to see the plant across the river as well as the rest of the Lehigh Valley. (Courtesy of Steelworkers Archives.)

Contemporary artwork added a welcoming touch to the lobby upon entering the glass doors. The Annex buildings and visitor parking lot are beyond the tall glass windows. (Courtesy of the National Museum of Industrial History.)

The Charles M. Schwab Memorial Library was on the lobby level of Martin Tower, opposite the main entrance. Employees were able to sit, relax and read a book, or conduct research. The library had an extensive collection of journals, manuals, and books. (Courtesy of the National Museum of Industrial History.)

Executives pose for a group photograph in the 21st-floor executive conference room. (Courtesy of the National Museum of Industrial History.)

The 21st-floor executive conference room offered a long oak table where critical decisions about the company were made. Adjoining the conference room was a kitchen where meals were prepared for the top executives. The conference room offered panoramic views of the Lehigh Valley for decades. (Courtesy of Steelworkers Archives.)

After Martin Tower and the Annex buildings were erected, the Stewart S. Court Auditorium was added to the Martin Tower complex. It was between the two Annex buildings and was a short walk from Martin Tower. It was larger than the auditorium at the old SGO building and had a more advanced lighting and sound system. (Above, courtesy of the Easton Area Public Library; below, courtesy of the National Museum of Industrial History.)

Seven
Revitalization, Preservation, and Future Plans

There are a few buildings on Bethlehem Steel's plant site that have been repurposed and spared from the wrecking ball. The National Museum of Industrial History is housed in the former electrical repair shop. This Smithsonian affiliate museum houses various pieces of machinery from the steel and textile industries that were once a huge part of the local economy. It opened its doors in 2016. (Author's collection.)

The Bethlehem Visitors Center is housed in the former Stock House that was built in 1860. Here, visitors will find interactive exhibits, including a film showcasing Bethlehem Steel as well as other local items of interest. Staff members are on hand to help answer questions and provide information on other attractions in the area. (Author's collection.)

The Lehigh Heavy Forge Corporation is the oldest operational building at the Bethlehem Steel site. It is well known for its field rings and generator shafts at the Niagara Falls power-generating plant as well as the 56-ton Ferris wheel axle used at the Chicago World's Fair in 1893. Along with its parent company, WHEMCO, which purchased the building in 1997, it continues to make forged steel components. It supplies metal, general, and ship-building industries throughout the world. Employees continue to find artifacts from Bethlehem Steel. (Courtesy of Jacob Matus.)

The Lehigh Valley Charter High School for the Arts relocated and opened its doors south of the site of the Alloy Mills in 2015. The school offers dance, theater, vocal/instrumental music, visual art, and figure skating. Multiple performances and art shows are exhibited throughout the year to showcase the students' talents. (Author's collection.)

The Steel Ice Center sits in what used to be the Alloy Mills. It broke ground in 2002 and opened to the public in 2003. The center offers figure, ice hockey, and recreational skating. Local schools also rent the facility for hockey games and tournaments. (Author's collection.)

Sands Casino was built where the ore yard used to be, off the Minsi Trail Bridge. It opened in 2009 and is the size of three football fields holding over 3,000 slot machines, 200 table games, and 36 poker tables. In 2020, Windcreek bought Sands. It has plans to invest $20 million into the adjacent No. 2 machine shop to make it an indoor water park and hotel. (Author's collection.)

The Sands Outlet Center consists of shops, a children's arcade, restaurants, and offices. The adjoining hotel offers 282 guest rooms, an indoor pool, fitness center, spa, business center, and conference facilities adjoining the casino and extending under and past the Minsi Trail Bridge. All the Windcreek facilities display prints of the former Bethlehem Steel and other landmarks in the Lehigh Valley. (Author's collection.)

The Turn & Grind shop (behind the visitors center) is a 26,000-square-foot building that was erected around 1890. In a January 2019 press conference, it was proposed as a multi-use facility for educational classes and exhibits as well as theater and dance performances. It is also planned to be used for Arts Quest's annual festivals, such as Musikfest, Christkindlmarkt, and Octoberfest as well as corporate and private events. (Author's collection.)

Central tool shop faces the Hoover-Mason Trestle with the SGO building in the background. Although the buildings are fairly intact and in good condition, the future of these iconic landmarks is unclear. (Author's collection.)

The No. 2 machine shop annex is one of the largest buildings still on the original Bethlehem Steel site. It was heavily used for manufacturing material to win the world wars. Many project proposals have been submitted to remodel and restore this brick and steel building with cast iron columns over the years, but so far, nothing is definite. (Author's collection.)

The No. 3 treatment building (also known as High House) sits along the Minsi Trail Bridge between the Lehigh River and Daly Avenue. Forgings such as a shaft or gun barrel were heat-treated and strengthened horizontally in these buildings to prevent sagging. A replica of a gun barrel produced in the treatment building is displayed outside the building. (Author's collection.)

Along the Hoover-Mason Trestle, there is a sky walkway that takes one from the steel stack blast furnaces (right) up to the Wind Creek Casino, outlet center, and tall hotel on the left below. It also goes past the No. 2 machine shop, which continues to be exposed to the elements (below). (Both, author's collection.)

The No. 6 ore trolley as well as the No. 2 ore trolley sit exposed to the elements along the Hoover-Mason Trestle. The No. 2 ore trolley is pictured on page 20 when it was in operation in the 1980s. Monthly tours are given by retired steelworkers and members of Steelworkers Archives along the trestle. (Both, author's collection.)

The iron foundry, also known as building No. 84 (above), and the Bessemer Rail Rolling Building (below) are some of the oldest buildings on the site of the Bethlehem Steel plant. The foundry has been used to store pallets for local festivals. At present, there are no preservation efforts planned, even though these buildings are over 100 years old. (Both, author's collection.)

Discover Thousands of Local History Books Featuring Millions of Vintage Images

Arcadia Publishing, the leading local history publisher in the United States, is committed to making history accessible and meaningful through publishing books that celebrate and preserve the heritage of America's people and places.

Find more books like this at
www.arcadiapublishing.com

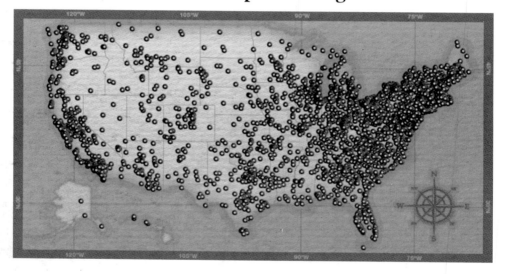

Search for your hometown history, your old stomping grounds, and even your favorite sports team.

Consistent with our mission to preserve history on a local level, this book was printed in South Carolina on American-made paper and manufactured entirely in the United States. Products carrying the accredited Forest Stewardship Council (FSC) label are printed on 100 percent FSC-certified paper.